Water III

转啊转

Keep on Turning and Turning

Gunter Pauli

［比］冈特·鲍利 著

［哥伦］凯瑟琳娜·巴赫 绘

郭光普 译

上海远东出版社

特别感谢以下热心人士对童书工作的支持：

匡志强　宋小华　解　东　厉　云　李　婧　庞英元
李　阳　梁婧婧　刘　丹　冯家宝　熊彩虹　罗淑怡
旷　婉　王靖雯　廖清州　王怡然　王　征　邵　杰
陈强林　陈　果　罗　佳　闫　艳　谢　露　张修博
陈梦竹　刘　灿　李　丹　郭　雯　戴　虹

目录

转啊转 4

你知道吗? 22

想一想 26

自己动手! 27

学科知识 28

情感智慧 29

艺术 29

思维拓展 30

动手能力 30

故事灵感来自 31

Contents

Keep on Turning and Turning 4

Did You Know? 22

Think About It 26

Do It Yourself! 27

Academic Knowledge 28

Emotional Intelligence 29

The Arts 29

Systems: Making the Connections 30

Capacity to Implement 30

This Fable Is Inspired by 31

一条鲑鱼用尽力气地逆流而上，希望尽快到达生宝宝的地方。一只蝴蝶在上方毫不费力地飞来飞去，看到水中的鲑鱼费力地大口喘气并奋力越过岩石。

蝴蝶问道："你不累吗？"

A salmon is swimming against the current with all its strength, hoping to get home to its spawning ground soon. A butterfly, moving effortlessly through the air, sees the salmon straining, gulping for air and leaping over the rocks.

"Are you not getting tired?" Butterfly asks.

一条鲑鱼逆流而上……

A salmon is swimming against the current ...

……河水总是在改变方向……

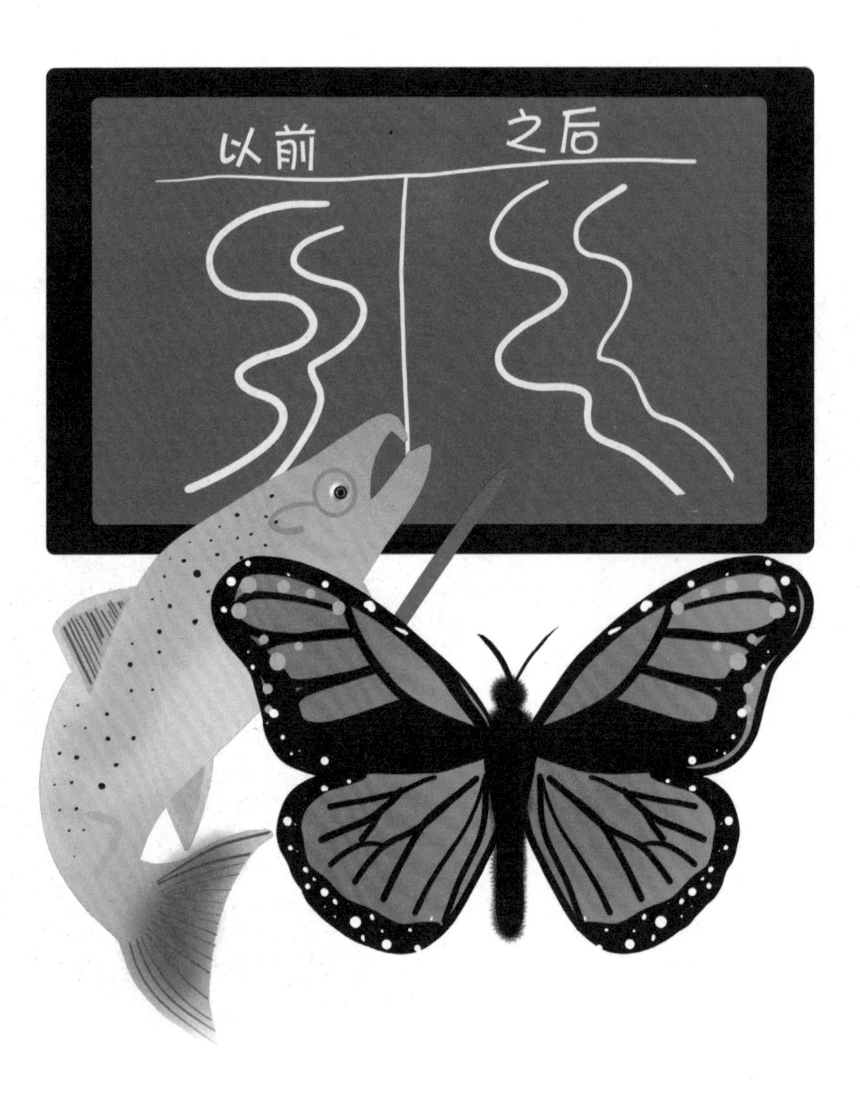

... river keeps on changing direction ...

“我不介意逆流而游，但是为什么河水总是在改变方向呢？”鲑鱼抱怨道。

“你是因为太累了才抱怨的吗？密西西比河里的小虾比你小多了，但他们可以轻松地游到离上游两倍远的地方。”

“I don’t mind swimming against the current, but why does the river have to keep on changing direction?” laments Salmon.

“Are you complaining because you are getting tired? The shrimp in the Mississippi River are smaller than you and they can effortlessly move twice as far up the river.”

“不，我不是累！拜托你不要把小虾和鲑鱼做比较。说真的，为什么河水不沿着直线流呢？这样我的活动就可以轻松5倍了。”

“噢，有人坚信在两点间的直线上运动是最省力的，这么说你也是他们中的一员喽？”蝴蝶问道。

“No, no, I am not tired! And don’t you compare shrimp to salmon! Seriously, why do rivers not flow in straight lines? That would make my work five times easier.”

“Ah, so you are one of those who believe that the least effort is required by moving in a straight line between two points?” Butterfly asks.

……直线运动……

… moving in a straight line …

……以为你喝醉了！

… thought you were drunk!

“好吧，我知道实际上你也不能沿直线飞行。我记得我的朋友鳟鱼说，当他第一次见你飞的时候，还以为你喝醉了！”鲑鱼大笑道。

“听着，我知道地球是圆的，风和水流使船不可能沿直线从一个地方航行到另一个地方，但是像船一样左右变换、前后回转地利用风向也可以让我很省力地向前飞。”

“Well, I know for a fact that you are incapable of flying in a straight line! I remember that my friend the trout, when he first saw you flying, thought you were drunk!” Salmon laughs.

“Look, I know the Earth is round, and that winds and currents make it impossible for a boat to sail in a straight line from one point to another, but turning back and forth from left to right against the wind as boats do, allows me to move forward with little effort.”

“你说的是可以自由活动的广阔大海。而我们鲑鱼所要面对的是一条蜿蜒曲折的河流。我都怀疑这些无尽的弯曲是否属于自然逻辑的一部分。”

“这证明了自然界存在逻辑。”蝴蝶回答道。

“我宁愿认为这是自然界缺乏逻辑的证明。”鲑鱼叹息道，“我希望下辈子做一名工程师。”

“Now you are referring to the open sea where one has the freedom to move around. What we salmon have to deal with here is a river, and one that is meandering. I wonder if these endless bends are part of Nature's logic.”

“It is proof of that very logic,” replies Butterfly.

“I rather think it is proof of the lack of logic in Nature,” sighs Salmon. “I think I'd like to be an engineer in my next life.”

……我希望做一名工程师……

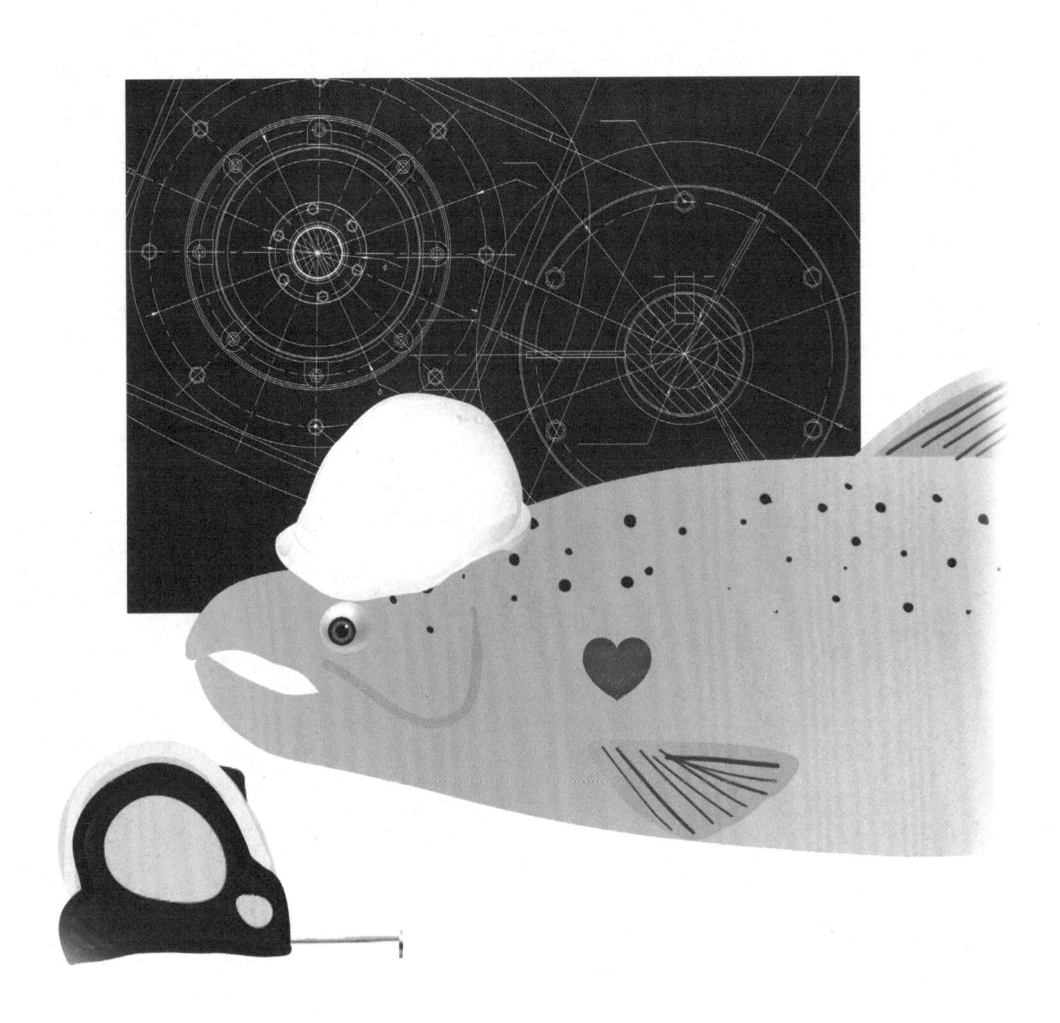

... I'd like to be an engineer ...

……更多的食物就会到达相应的地方。

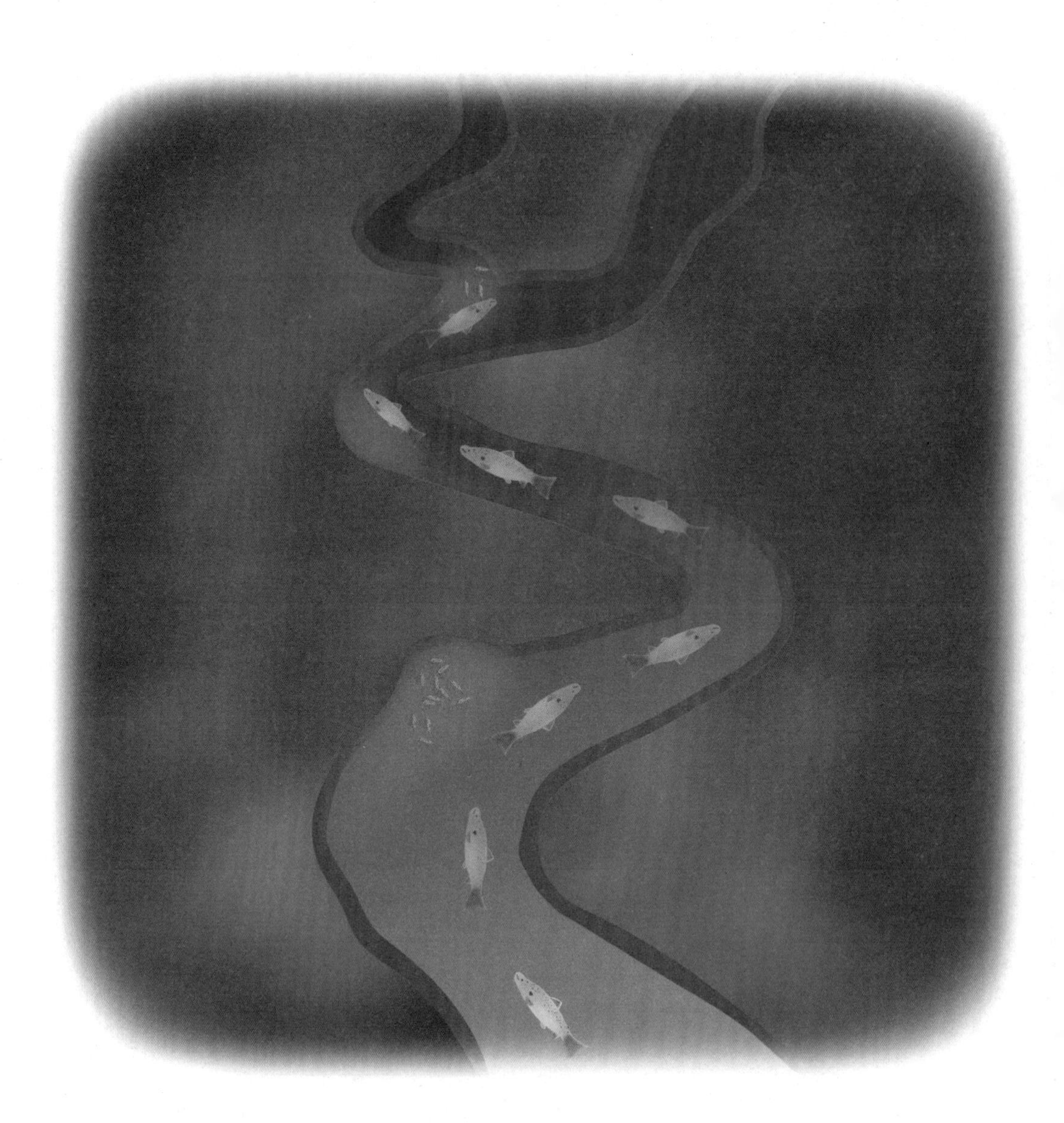

… more food gets to where it needs to go.

“这不是一种很礼貌的对待自然——我们生命之母——的方式。如果河流没有弯曲，只能让生活更加艰难；每次改变方向，更多的食物就会到达相应的地方。”

“你的意思是说，河流侵蚀河岸形成曲线，沙子会在河中央沉降下来，这是一种获取食物的方式吗？谁吃石头和沙子啊？”

“That is not a very polite way to refer to Nature, our Mother of Life. Rivers don’t have bends just to make life harder for you; with every change of direction more food gets to where it needs to go.”

“You mean that water digging into the river banks where they curve and the sand dropping down to the centre of the river, is a way to get food? Who eats rocks and sand?”

“你不能把一条河看成仅仅是运输石头和沙子的工具。人们也不应该把水看作是运输垃圾的工具。河流里有那么多的生命，而所有的湍流都能使水中的氧气丰富起来！”蝴蝶解释道。

“生命？什么生命？我并没有在河流和小溪中看到许多生命。”

“You cannot reduce a river to just a transporter of rocks and sand. Neither should people consider water as a transporter of their waste. There is so much life in a river, and all that turbulence enriches the water with oxygen!” Butterfly explains.

“Life? What life? I do not see a lot of life in these rivers and creeks.”

……湍流能使水中的氧气丰富起来！

... turbulence enriches the water with oxygen!

……河岸……

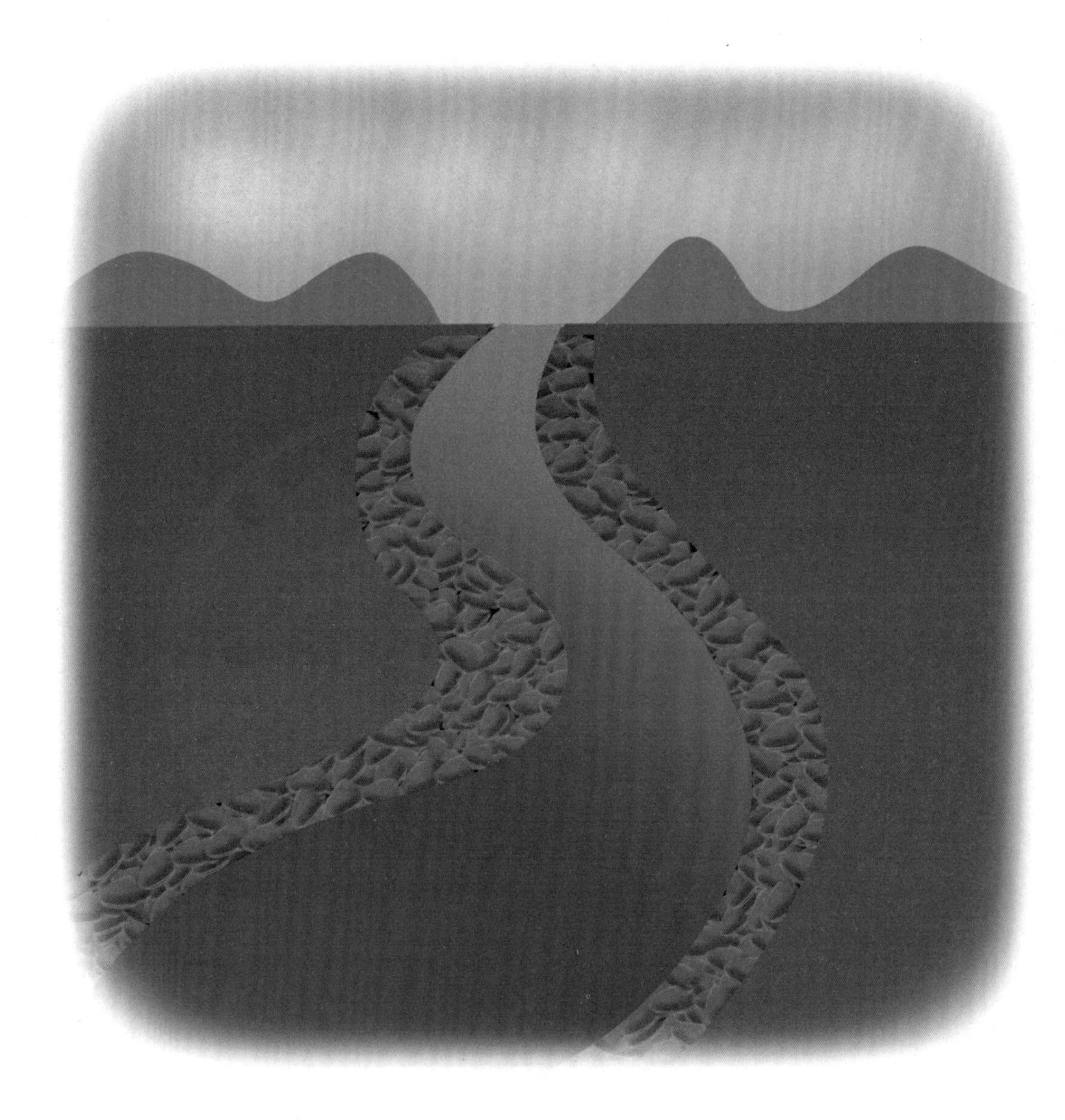

… river banks …

“嗯，你需要放慢速度，仔细看看。河流为一个地区补充地下水——这样可以使水不会流得太快或者直接流走。还有食物以浮游生物的形式黏附在每一块岩石上。”

“那么为什么有人要建混凝土河岸，强迫水直线流淌，而不是让水按照自然的方式流淌呢？”

“Well, then you need to slow down and take a closer look. Rivers replenish the area with groundwater – providing they don’t flow too fast or in a straight line. And there is food in the form of plankton sticking to every rock.”

“So then why have people been building concrete river banks, forcing the water to flow in a straight line, and preventing the river from running the way Nature intended?”

“噢，无论什么，只要是以错误的方式建立起来的，就肯定需要重建，”蝴蝶说，“以便把物理定律考虑在内，防止无情的水流把一切冲毁。”

“这是不是意味着人们应该像我这样利用水的力量？”

“是的，你是利用各种物理力量为自己服务的专家。为此，我们赞美你！”蝴蝶高兴地说。

……这仅仅是开始！……

“Well, whatever has been built in the wrong way will simply have to be rebuilt,” Butterfly says,“to take the laws of physics into account, and prevent the unrelenting movement of water from knocking everything out of place.”

“That means people should use the power of water like I do?”

“Yes, you are a master of using every physical force to your advantage. And we celebrate you for that!” Butterfly exclaims with joy.

... AND IT HAS ONLY JUST BEGUN! ...

……这仅仅是开始！……

… AND IT HAS ONLY JUST BEGUN! …

Did You Know?

你知道吗？

It takes a little disturbance and a lot of time for a river that runs in a straight line to turn into one that curves. In Nature there is a lot of both.

要让一条直的河流变成一条弯曲的河流，需要的是少量干扰和大量时间。在自然界中这两种条件都很充足。

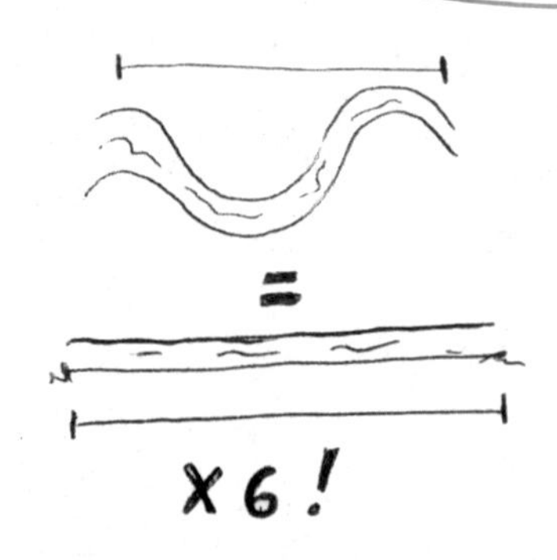

The length of a complete S-curve of a river tends to be six times the straight of the stream. The collection of S-curves of creeks and rivers is asymmetrical, and from the air looks like fractals.

一条S形河流的长度可以是直线形的六倍。把许多S形溪流和河流汇聚在一起，会得到一个不对称图形，从空中看有点像分形几何。

An oxbow lake is created when two S-shaped curves dig deeper and deeper and after thousands of years connect—creating a short cut that leaves behind a part of the river.

当两条S形河流的河床被冲积得越来越深，数千年后连通到一起时就会形成牛轭湖——在河流后部会出现一段直的河道。

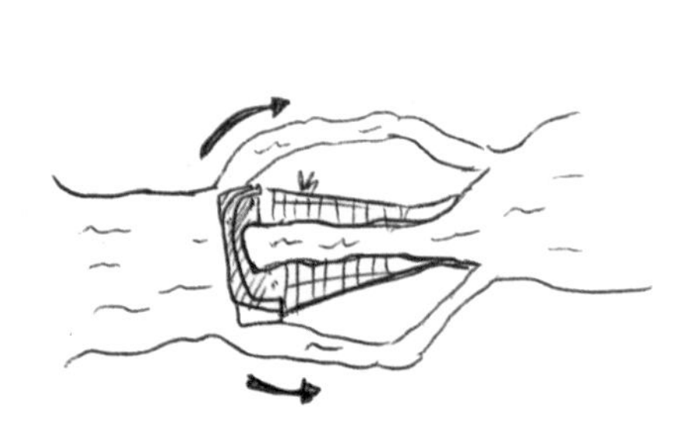

Rivers are known as "self-organising" and whenever engineers attempt to block erosion, or change the direction of a river, it will always evolve back to its natural optimum.

河流广为人知的一个特点就是"自组织"，并且无论工程师在何时试图阻止河道侵蚀或者改变河流流向，它总会演化回到自然条件下最适宜的形态。

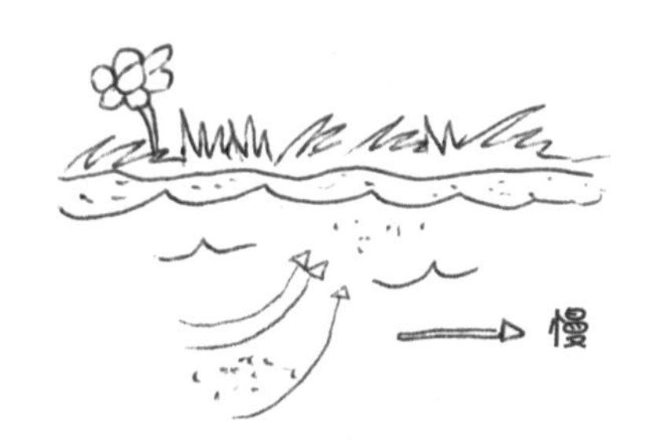

Water flows more slowly in shallow areas near the inside of a river bend. Slower water can't carry as much sediment and deposits its load in the riverbed.

水在河湾里的浅水处流动得慢一些。较慢的水流不能携带大量物质，于是沉积物沉积在河床上。

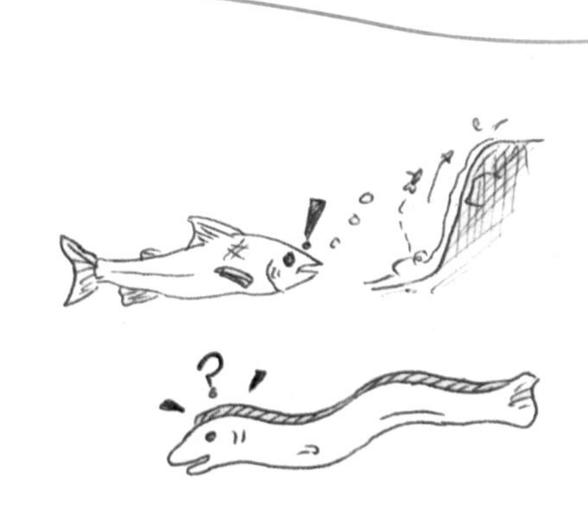

When people "tame" a river, they stop the transport of sediment, which is a nutrient for soil. This affects the movement of salmon and shrimps swimming upstream and of eel swimming downstream.

当人们“驯服”了一条河流，他们就阻止了沉积物的运输，而沉积物正是土壤肥力的来源。这就影响了鲑鱼和小虾游向上游以及鳗鱼游向下游。

The silt that rivers create and carry, maintains deltas that build landmass. This prevents coastal zones from being eroded by the incessant ebb and flow of the ocean. It also protects against rising sea levels as a result of climate change.

河流产生和携带的淤泥维持了构成大陆主体的三角洲。这就避免了海岸区域不断被潮水和洋流侵蚀。这也预防了由于气候变化导致的海平面上升。

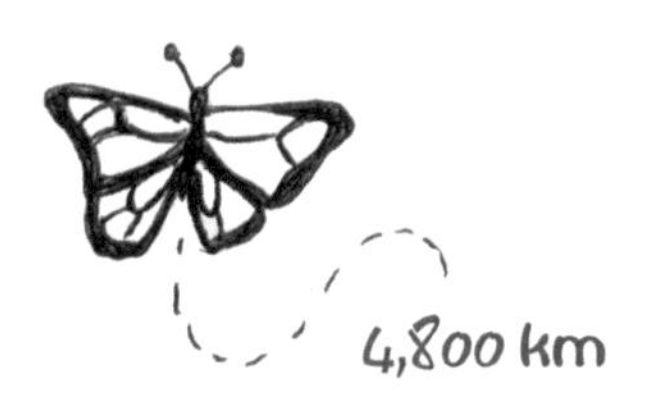

A juvenile shrimp, no bigger than a 6 mm can crawl and swim up to 1,600 kilometres upstream against strong currents – that is even further than a salmon can swim. The monarch butterfly migrates for distances of up to 4,800 kilometres.

一只幼年小虾，不到 6 毫米长，可以逆着湍急的水流爬行和游行 1 600 千米，甚至比鲑鱼游得还远。黑脉金斑蝶迁徙的距离可长达 4 800 千米。

Think About It
想一想

Is a meandering river energy efficient, or does it cause too much trouble for anyone using the river?

弯曲的河流是一种高效能量吗？还是说它会给任何一个依靠河流生活的人带来很多麻烦？

What makes salmon and shrimp want to travel such long distances, just to have offspring?

是什么导致鲑鱼和小虾想要游这么远的距离？只是为了繁衍后代吗？

What is the difference between Nature's Logic and the logic of an engineer?

自然逻辑和工程师逻辑的区别是什么？

Would you want to straighten a river, or let it find its own way?

你想让河流变直吗？还是顺其自然？

Do It Yourself!

自己动手！

Let's take a look at a meandering river. Is there one close to where you live? Is this river flowing in a straight line, as engineered by humans, or meandering? See if you can find an aerial photograph, or look on Google Earth. Take a measurement of a section that includes at least 10 meanders. Calculate the total length of each of these meanders, following the completion of the S-shape. Measure along the curves. Now measure the width of the river. What is the ratio between width and length? The information you have now gathered gives the basic number that allows you to produce fractals.

让我们来看一看弯曲的河流。你住的附近有河流吗？这是一条像人类设计建造的直线河流还是弯曲的河流？看看你能不能找到河流的航拍图或者在谷歌地图上看到。测量至少包含10个弯道的一段河流。测量S形的长度，计算弯道的总长度。再测量河流的宽度。河流的宽长比是多少？根据你搜集的基本数据信息可以得出一些分形。

学科知识

Academic Knowledge

生物学	鲑鱼溯河产卵，这意味着它们出生在淡水，之后都生活在咸水中，在生命的最后它们洄游到河流中，在砾石河床上产卵后便死去；砾石上有丰富的浮游生物，可作为鲑鱼产卵的床，为孵化出来的小鱼提供第一顿大餐；幼鲑鱼或其他鱼类的生理变化将帮助鱼类适应从淡水到咸水的转换。
化　学	鲑鱼尸体的分解可以把重要的营养成分释放到环境中，把海洋中的氮、硫、碳和磷补充到陆地；当鲑鱼卵孵化成小鲑鱼时，化学线索就烙印在小鱼身上，这可以引导鲑鱼回到自己的出生地。
物　理	河流内弯道的压力与外弯道的离心力平衡；蜿蜒的河流下游流体在外弯道速度更高；由于角动量守恒，内、外弯道的速度不同。
工程学	使用鱼梯或鱼道，为迁徙的鱼类提供绕过河流中的特殊障碍物的路线；河流工程师是一种职业。
经济学	驯服河流造成生态系统服务功能的损失；考虑运输时间上的生产力，忽略了有更多附加值的整个生态系统的再生能力。
伦理学	把河流变直以及硬化河床会丧失许多生态系统功能，随着时间的推移将导致洪水暴发、生物多样性减少、地下水补充不足。
历　史	在古希腊时代，有一条河被称为“蜿蜒河”，今位于土耳其境内，“蜿蜒”这个词，指曲折缠绕的事物，就是源于这条河的名字。
地　理	沿河流方向最大水深处的连线称为深泓线；当河流用于政治界限时特指边界。
数　学	高压导致低速度；波形蜿蜒的河流沿着山谷的轴线，这条直线使曲线由此测量的所有振幅的总和为零；在现实中，因为地球的形状和风向，最短的距离并不是一条直线。
生活方式	“顺其自然”是一个受到水流启发产生的成语，水总能找到阻力最小的路径前行。
社会学	牛轭湖在不同的语言中有各种各样的名字，荷兰称“马蹄湖”，德国称“老水湖”，法国称“死亡之臂”，澳大利亚称“死水潭”。
心理学	在自然界中需要非凡的努力和决心才能繁殖成功，从婴儿的生命之旅，到鱼儿逆着或顺着河流的长途旅行以及蝴蝶和鸟类的迁徙。
系统论	人类通过最短的距离和最快的时间追求效率,但是大自然的效率是基于同时考虑到多种功能和职责。

情感智慧

Emotional Intelligence

蝴　蝶

蝴蝶对鲑鱼表示同情。她对他的生活感到担忧，把摆在他面前的任务和别人的相比，言明别人的挑战更大，以此来鼓励他。蝴蝶质疑鲑鱼的逻辑，这表明她对友谊充满信心。蝴蝶用清晰的逻辑来解释她的推理方式。当鲑鱼提出质疑时，蝴蝶只是简单地做了个声明，然后停止争论。蝴蝶从讲逻辑转变为讲目的，指出整个系统的收获，而不仅仅是一个单独的功能。蝴蝶让鲑鱼要有耐心，并从另一角度看问题，她质疑他接受人类使用的逻辑。通过赞美鲑鱼的成就，蝴蝶加强了他面对困难时期的决心。

鲑　鱼

鲑鱼对他与蝴蝶的关系有信心，这使他能自在地向她抱怨。然而，鲑鱼不希望被认为是软弱的，他在寻找高效前进的方法。鲑鱼坚持认为海洋和河流的状况是有区别的，并用人们通过破坏性的工程来改变河流的方式这个案例来寻求支持。但随后他认识到蝴蝶的观点是正确的，便与蝴蝶的逻辑达成一致，这是基于他利用水的力量的能力。

艺术

The Arts

让我们来看看分形世界，大自然如何用独特的方式来建构自己。轮廓和形状是美丽的——然而，潜在的图案遵循精确的几何逻辑。在大自然中寻找各种各样的分形，欣赏它们的美。现在做一些研究,看看你能否找到有关蜿蜒曲折的河流的航拍或无人机拍的照片，例如在南美洲亚马孙河的河口、流经沙漠的河流，还有纳米比亚的鱼河峡谷。讨论它们的美丽，看看你是否可以用不同的艺术材料重塑它们的鸟瞰图。

思维拓展

Systems: Making the Connections

工程世界追求的是效率——只用时间和成本作为衡量参数。尽管这种逻辑在追求生产率的社会中盛行，但这并不是大自然的逻辑。大自然的逻辑是处于一种永恒的、源源不断的活动状态，以满足每个人的基本需求。很明显，工程师的逻辑与自然的逻辑是不同的。工程师可能认为河流截弯取直将节省货物的运输时间并减少能源消耗，而忽略了一个事实，即除了考虑时间和成本之外，还有其他目标要满足。要使生命茁壮成长有几个条件，如作为营养的矿物质的供给，漩涡丰富了水的含氧量，三角洲的存在能保护堤岸免受潮水涨落的冲击，还有流速缓慢的河流能补充地下水。此外，在自然界的定律作用下，河流变成了自组织系统，蜿蜒是由于压力差异而导致的，并产生了一个规则，干扰出现时，这个规则总是起作用并获胜。蜿蜒的河流不仅服从物理定律，也是一种生命的力量，为各种各样的生物以及河流流域提供能源（氧气）、水和营养。这就形成了应对同样来自自然的力量的物理保护，如潮涨潮落的冲击、由于气候变化导致的海平面上升，以及风暴对海岸线的冲击。河流的曲折制造了一种物理屏障，保护了海洋和陆地之间脆弱的交界带，这个地区是生命和能量被创造和边界得以维持的界限。在这种前提下，迁徙的鱼类战胜了一个又一个挑战，它促成了“海洋—陆地”关联，给淡水域带来了丰富的营养，或通过捕食迁移，或通过分解鲑鱼产卵后的死尸。这个能量传递保证了很少被认识到其重要性的生命循环。

动手能力

Capacity to Implement

绘制一幅关于流经季节性河流的水量意境地图。如果由于河道的矫直而水流迅速，并且有坚固的钢筋混凝土建造的堤岸保护，那么水就没有时间渗出并补充地下水了。如果建设淤地坝，在河里创造储水区，那么水就可以长时间保留在河道中并足以穿透地面！现在，你可以主张建设淤地坝来帮助农业，因为它增加了可用水量，而不需要任何费用。

故事灵感来自

This Fable Is Inspired by

阿斯霍克·寇斯勒

Ashok Khosla

阿斯霍克·寇斯勒，联合国环境规划署前署长，他的理念包括了源自甘地的创造自给自足的基于生物资源发展的村庄共同体概念。寇斯勒博士是一名物理学教授，在哈佛大学任教，也曾为联合国工作。他得出结论：水资源短缺、森林砍伐、水土流失、人口爆炸是导致印度本德尔肯德地区农业产量低下的主要原因。作为回应，他的组织设计了一种替代发展方案，开始将淤地坝作为一种适当的干预措施，以恢复地下水，否则河水将由于直流河道而不再（向土地中）渗水。通过这种方式，他帮助当地居民满足了他们的基本需求。自 1989 年以来，替代发展方案推动了印度本德尔肯德地区近 100 座淤地坝的建设。

图书在版编目(CIP)数据

冈特生态童书.第四辑:修订版:全36册:汉英对照/
(比)冈特·鲍利著;(哥伦)凯瑟琳娜·巴赫绘;
何家振等译.—上海:上海远东出版社,2023
书名原文:Gunter's Fables
ISBN 978-7-5476-1931-5

Ⅰ.①冈… Ⅱ.①冈… ②凯… ③何… Ⅲ.①生态环
境-环境保护-儿童读物—汉、英 Ⅳ.①X171.1-49

中国国家版本馆CIP数据核字(2023)第120983号

著作权合同登记号图字09-2023-0612号

策　　划 张　蓉
责任编辑 张君钦
封面设计 魏　来李　廉

冈特生态童书
转啊转
[比]冈特·鲍利　著
[哥伦]凯瑟琳娜·巴赫　绘
郭光普　译